THE CRAZY APE

THE CRAZY APE

Written by a Biologist for the Young

by

Albert Szent-Györgyi

M.D., Ph.D., Nobel Laureate

Philosophical Library

New York

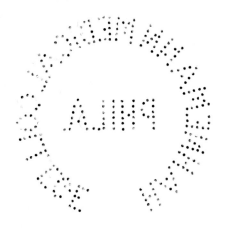

Manufactured in the United States of America

CONTENTS

THE CRAZY APE

FOREWORD

There can be little doubt that mankind is going through the most critical period of its history, a period which can very well end with his extinction in the not too distant future. Countless essays have been written about the causes and cures of this crisis. It has been analyzed from military, political, social, economic, technical and historical points of view. There is one factor, however, that has been largely forgotten: man himself—man as a biological entity. As a biologist, I intend to take this approach in this short book.

Mankind grows like a snake, shedding his skin now and then and acquiring a new skin. This process seems to coincide with the cycle of turbulence and quietude in man's history.

The sage of the Renaissance, Erasmus, distinguished between the quiet and the turbulent periods of history. The turbulent periods were those of sharp transition. The sharper the transition, the greater the tumult. So we have two questions to answer: what creates the sharp transition that is going on today, and how can man be fitted into his new skin? And, of course, the ultimate question: will mankind be able to survive the machinations of present-day men who often appear to act more like crazy apes than sane human beings?

9

I. THE PROBLEM IS STATED

Why does man behave like a perfect idiot? This is the problem I wish to deal with. Today is the first time in man's history that he is able to truly enjoy life, free of cold, hunger and disease. It is the first time he is able to satisfy all his basic needs. Conversely, it is also the first time in his history that man has the capability of exterminating himself in one blow or making his tragically shrinking, lovely little globe uninhabitable by pollution or overpopulation.

One would expect that any idiot could make a wise choice between these two alternatives. It is basically a choice between pleasure and pain. Yet man seems to be bent on choosing the latter, on bringing about the "Kingdom of Cockroaches." Cockroaches are very insensitive to high energy radiation and will find plenty to feed on in a world made barren of the resources needed to sustain human life. In the most affluent country of the world, five per cent of the people are starving. Fifty per cent are starving in the rest of the world—children do not get enough food to build healthy minds and bodies, and go to bed hungry. While this is going on the United States alone has spent since the Second World War a trillion dollars ($1,000,000,000,000) on "defense," on instruments of mass killing. The Soviets, of course, are not lagging far behind. These are sums too great to be visualized even by the keenest imaginations. They could long ago have lifted mankind to much higher levels of existence. This is a truly criminal story, yet it is not merely criminal. It is utterly stupid too, for what we have bought with all this paranoiac spending are insecurity, jitters, and a ticket for self-extinction, having placed our fate in hands that we have absolutely no reason to trust.

If mankind is really this idiotic, then how did it remain alive through its first million years? There are two possible answers to this question. One is that man is not really that senseless; rather, his circumstances have changed in such a way that he no more fits into his environment, which makes his actions senseless. But one could also conclude that man has always been like he is today, that he has always been self-destructive; he merely lacked the technological means to exterminate himself. Throughout his history there has always been an enormous amount of senseless killing and destruction; if man did not exterminate himself it was only because of the primitive nature and inefficiency of his instruments of killing, which ensured that in any violent conflict many would survive. Modern science has changed this situation, and today we all have to go together.

Whichever of the two theories is correct, if we are to have any hope of surviving it is most urgent that we find out what keeps us in this fateful groove and whether there is a chance and a way of getting out of it.

II. MAN vs. NATURE

Nature is big, man is small; the quality and level of human life has always depended on man's relation to nature, on the measure in which he is able to understand nature and use its forces for his advantage.

The survival of any species depends on its ability to adapt to its surroundings. Man is like any other species in that he has had to adapt himself to the world in which he was born. This world of, say, a hundred thousand years ago, was exceedingly simple, and so were its problems. The main problem was how to get through the day alive—find food, shelter, a sexual partner, whatever simple requirements were needed. So man developed senses which enabled him to distinguish between a bear or a wolf, rocks and trees, all the basic units of his world. His life improved at the rate with which he learned to shape and use things. The discovery of the needle, the wheel, the arrow, fire, metals, the hardening of clay, etc., marked the single stations on his upward road from primitiveness.

These discoveries were based on the everyday experience of man. It was not until that singular upsurge of the human intellect, characterized by ancient Egypt and the Greco-Roman world, that here and there men began to try to understand nature. These efforts can be summed up as the "science of antiquity." What was characteristic, on the whole, of the science of this period was a faith in the supreme power of the mind, which, it was thought, could solve all problems. This may be illustrated by the story of the two stones. Aristotle, whose ideas were the last word for many centuries to come, claimed that a big stone falls faster than a small one. What was remarkable about this statement is not that it was wrong, but that it never occurred to Aristotle to test it. He probably would have taken such a suggestion as an insult. Why resort to crude action when the mind could answer all questions? The freedom of human thought is very limited. We all live in a very narrow cage, the "spirit of our times," in which we have very little freedom of motion. If, in different ages, people thought differently, this was not because the cage got wider, but because the cage moved. The spirit of his time made it impossible for Aristotle to pick up two stones to see which one fell faster.

In the 16th Century a great change must have taken place in the human mind for, one day, a querulous young man went up the Leaning Tower of Pisa carrying two stones, a big and a small one, to drop them simultaneously, having asked his companions to observe which of the two hit the pavement first. This man, Galileo Galilei by name, distrusted not only the perfection of his mind but also that of his senses, which he improved by building telescopes. By so doing he discovered the satellites of Jupiter, never seen by man before, proving that the universe could not have been built solely for man's pleasure or temptation. This rebirth of the human mind is what is summed up now as the "renaissance." Galileo was one of the first visionaries, followed soon by Kepler, Leeuwenhoek, and many others who measured, observed and calculated, building a classical science that reached its peak in Newton, Darwin and Pasteur.

This classical science dealt with the world as man knew it, in which man was born, to which he tried to adapt, in which he lived. Accordingly, this science brought no qualitatively new factors into man's life; it only cleared the inner relations of the surrounding world. This science had an enormous influence on human thought, replacing divine whim by law and order, giving man for the first time an idea of where he was and what he was.

While the science of antiquity did not change human life, classical science led, in the 19th Century, after a latent period of several hundred years, to the "industrial revolution," which greatly improved the quality of human life. It *improved* it, but brought no qualitatively new factors into it. The needle was known for thousands of years before; the sewing machine could only sew faster and better. Similarly, the "iron horse," the train, could outrun the real horse, and its invention made travel more comfortable. Death rates went down, production of food and commodities went up and a new social class, that of the industrial worker, was born; but on the whole the world structure remained unchanged.

At the turn of this century four important discoveries were made which marked the beginning of a new period in man's history. X-rays (1895), the electron (1895), radioactivity (1896) and the quantum (1900) were discovered, these discoveries being followed soon by relativity (1905). None of these were, or could be, revealed by our senses. They meant that surrounding man there was a world of which he had no inkling before, about which his senses could give him no information. They not only did not inform him, but specifically existed so as *not to inform him.* If they had they would have been useless and we would have had to die. If I perceived atoms or quanta instead of trucks, I would be run over; if my ancestors had seen electrons instead of bears, they would most likely have been eaten.

The story of man consists actually of two parts, divided by the appearance of *modern science* at the turn of this century. In the first period, man lived in a world in which his species was born and to which he and his senses were adapted. In the second, man stepped into a new, cosmic world to which he was a complete stranger. Never before in his history did he have to undergo such an acute transition. I am not so terribly old, but I can still remember the time my uncle, a scientist, told me that a paper was presented at the French Scientific Academy in Paris which proved definitely that flying with heavier-than-air objects was impossible. Everybody felt relieved because the idea of flying had begun to bother people. I can also remember the time the first car visited my father's farm and the farmhands demanded that the hood be opened and the swindle —that is, the hidden horse—be exposed.

After a latency period of no more than half a century, modern science began to transform human life, bringing factors into it about which man could not even dream before. The forces at man's disposal were no longer terrestrial forces, of human dimension, but were cosmic forces, the forces which shaped the universe. The few hundred fahrenheit degrees of our flimsy terrestrial fires were exchanged for the ten millions of degrees of the atomic reactions which heat the sun. The speed of the horse as a factor in human life was replaced by the speed of light or sound; the relatively inefficient force of our weapons was replaced by the forces of the atom, which can dig harbors, move mountains, or annihilate whole societies in seconds.

John Platt* summed up the change by showing that in this century we have increased our speed of communication by a factor of 10^7 (ten million fold), our speed of travel by a factor of 10^2, our speed of data handling by 10^6, our energy resources by 10^3, our power of weapons by 10^6, our ability to control disease by something like 10^2, and the rate of population growth to 10^3 times what it was a few

---

* **Science,** 166, 1115, 1969.

thousand years ago. This is but a beginning, and means
endless possibilities in both directions—a building of a
human life with undreamt of wealth and dignity, or a sudden
end in utmost misery. We live in a new cosmic world which
man was not made for. His survival now depends on how
well and how fast he can adapt himself to it, rebuilding all
his ideas, all his social and economic and political structures.
His existence depends on the question of whether he can
adapt himself faster than the hostile forces can destroy him.
At present, he is clearly losing out.

We are forced to face this situation with our caveman's
brain, a brain that has not changed much since it was
formed. We face it with our outdated thinking, institutions and methods, with political leaders who have their
roots in the old, prescientific world and think the only
way to solve these formidable problems is by trickery and
double talk, by increasing our atomic arsenal—which is
already sufficiently stocked to kill every single living individual three times over—by trying to replace the single war-
head by multiple ones, creating new missiles and anti- or
anti-antimissiles, by spending untold billions on the instruments of death. We can already wipe out any distant city
in one blow; nevertheless we are still placing more and
more of these bombs under the ground and under the sea,
ready to be fired, as if they were bombs of the old type,
the number of which could decide battles.

What makes all this so terribly senseless is that these
bombs cannot be used at all. They are too powerful. We
cannot fire them without committing suicide and wiping
out mankind altogether. The world's biggest military power
is at present unable to cope with a small and undeveloped
nation, a nation that has none of these bombs and yet
drains the vital forces of its opponent. A world in which I
can watch in living color my fellow man stepping on to

17

the moon, and even hear him talking, sitting myself in slippers at home, is not the old world in which mankind was born. It is a new world, and it demands new ideas, leaders and methods. That we have not yet come to this realization—that we have conceived no "new" ideas, have developed no "new" leaders, have devised no "new" methods—is made depressingly obvious by the fact that we are still acting like man of thousands of years ago. Through the ages, man's main concern was life after death. Today, for the first time, we find we must ask questions about whether there will be life before death.

III. THE BRAIN AND THE MIND

Whatever man does he has to do it first in his mind, and the mechanism underlying the mind is the brain. There can be no action without an underlying mechanism, and a mechanism can only do what its structure allows it to do. A cow could never lay an egg, however hard she tried; nor could a gramophone type letters or a typewriter make music. Man, too, can do only what his brain allows him to do. Thus, when discussing human action we must have a look at the brain and see what sort of organ it is and for what purpose it was shaped by nature.

In their struggle for life some animals grew fangs, others claws or tusks, while still others produced poisons. Man grew a brain, and it is a curious fact that this semisolid blob of matter proved to be a more formidable tool than fangs, claws or poisons, and insured man's supremacy. Man's brain was not developed by nature to search for truth, but to search for food, safety, and the like; to search for advantage, to help man get through the day alive. It is an organ of survival. Human action is motivated by need or desire, and the brain is the instrument of human gratification.

In primitive societies this must have been all there was to it. In more sophisticated societies the brain developed a second function: to find arguments, mostly high-sounding ones, to justify deeds or desires. This our brain does so promptly that we kid ourselves into believing that we are actually motivated in our actions by these arguments.

Talking about arguments, one point must be made clear: they have no real meaning. They consist of words, and words can be put together in many ways. Everybody knows about the favorite pastime of Socrates: "Say something, and I will disprove it; then say the opposite and I will disprove that too." Anything can be justified by words and logic.

Our nation has lately been polarized into two sets of attitudes—those of hawks and doves. The first like war, the second like peace. Both can justify their attitudes and actions equally by words and logic. So these have no real meaning. What has meaning is the blunt fact that there are hawks and there are doves, that *hawks think and act like hawks, doves think and act like doves*.

Our whole nervous system developed for one sole purpose, to maintain our lives and satisfy our needs. All our reflexes serve this purpose. This makes us utterly egotistic. With rare exceptions people are really interested in one thing only: themselves. Everybody, by necessity, is the center of his own universe.

When the human brain took its final shape, say, 100,000 years ago, problems and solutions must have been exceedingly simple. There were no long-range problems and man had to grab any immediate advantage. The world has changed but we are still willing to sell more distant vital interests for some minor immediate gains. Our military-industrial complex, which endangers the future of mankind, to a great extent owes its stability to the fact that so many people depend on it for their living.

This holds true for all of us, including myself. When I received the Nobel Prize, the only big lump sum of money I have ever seen, I had to do something with it. The easiest way to drop this hot potato was to invest it, to buy shares. I knew World War II was coming and I was afraid that if I had shares which rise in case of war, I would wish for war. So I asked my agent to buy shares which go down in the event of war. This he did. I lost my money and saved my soul.

These traits of the human mind have remarkable consequences for social structures. Man creates institutions to satisfy his social needs in accordance with his philosophy. Individuals join these institutions and make their personal interests fuse with those of the institutions, on whose wealth and power their own prospects depend. What follows is that very soon these institutions begin to serve their own interests rather than social needs. As time goes by the social needs and philosophy change, but the institutions don't; they remain fighting for their own interests until they are swept away by revolution, often at the price of much suffering, bloodshed and devastation. Man grows by virtue of these upheavals; he becomes, then, like a snake, bursting his skin periodically.

The emergence of modern science has greatly speeded up these changes. Most of our social institutions now serve mainly their own interests while pretending to serve the purpose for which they were created. This holds equally for armies, churches or governments and means that we are living in a hypocritical world, one of false pretenses, one which is now being rejected wholesale by our youth.

IV. REMARKS ON EDUCATION

I have tried to show in the previous pages that function depends on an underlying mechanism and that human action has its foundation in the structure of the brain. If this would be all there is to it, nothing could be done to change the present course of things, and this book would be superfluous. What places man's future into his own hands is the fact that his brain has many traits in common with a machine invented in the last decade: the computer. The computer is a mechanism like any other, but what it can do depends not only on its coarse structure but also on its finer electric or magnetic structure, which can be changed by programming. What a computer can do, then depends on its programming.

What we call "education" is nothing but the programming of the brain at an early stage when it is still malleable. The future of mankind depends on education, a system of programming which can be changed. Human history reflects, in essence, the gradual change in this programming, and if you compare yourself with a savage cannibal you may find that the only essential difference between the two of you is in the educational programming the two of you underwent. It follows from this that education is one of the most important activities of mankind. It opens the door to wonderful possibilities, but it also exposes mankind to terrific dangers, for any dictatorial system can, through education, transform society according to its interest and, if desired, can transform decent people into savage killers, as we have seen happen repeatedly within the present century.

Political systems have invariably availed themselves of these possibilities. The first task of any new political system has always been to create an educational system that will serve its interests and secure its stabilization. I myself grew up in a feudalistic country where thinking was considered dangerous; thus, I was made to study endlessly and was taught that the highest virtue is one's readiness to die for the king. (Later I learned that "king" actually meant the clique which started World War I with falsified dispatches presented to the Emperor.)

Theoretically the possibility exists, through education, of changing the course of history with one stroke, replacing today's narrow nationalism by human solidarity. Practically, however, there are enormous difficulties, for who is capable of teaching the young? It is the older people who have to teach the young, but older people tend to transmit to their students the world in which they have grown up themselves. Who should teach the teachers, then, and what sort of world is it we are aiming at? If we could answer these questions we would be half way on the road to solving the problems of the world.

All the same, there are approaches. One of the most important factors deciding man's actions is his system of values, most of which is instilled at an early age. If I, at my 76 years of age, am still impatiently running every morning to my laboratory, this is because as a child I learned from my family that the only thing worth striving for is the creation of new knowledge or beauty.

The subject which has the greatest influence on the formation of our system of values is history, for what other ground is there to build the future on than the past? Although I am a scientist, I think that the most important subject in a child's education is history. But it should be real history. "Real history," to my mind, is the story of the slow evolution of man, the story of how he rose from the level of his fellow animals to his present superior position, where he can appreciate beauty and knowledge and can think about where he is and what he is. The pace of this ascent is the result of two factors, one pulling it forward and the other pushing it back. What pushed it back were wars, bloodshed and destruction, the representatives of these forces being kings, barons, dictators, generals and the like. This is not only a bloody history, it is a false one too, for, as has been often shown, most wars were not decided by glorious kings or heroic generals but by rats and lice which carried infections and spread plagues. They were also decided by nutritional deficiencies, an example of which is the recent Biafra tragedy. The representatives of the forces which propelled man forward were those who searched for and found new knowledge and beauty, more cogent ethical and moral values. Yet I found their names mentioned nowhere in my history books. The height of the pedestals of the statues we erect to our national heroes is mostly proportional to the number of people they killed, as Bertrand Russell remarked. To my mind the real heroes of mankind are the Galileos, Newtons, Darwins, Pasteurs, Shakespeares, Bachs, Lao Tzus and Buddhas, whose names one seldom sees noted in history books filled with descriptions of battles and meaningless shifts of national borders.

A good education could also solve another of the most pressing problems of man: what to do with himself when he has reached the point that he can produce more than he can consume. For this we need an educational system based on the real understanding of moral, esthetic and spiritual values. There is enough beauty, grace and greatness in the world to fill the mind; there is no real need to go out and kill people in order to shake off boredom.

The situation described in this chapter has changed lately to some extent. I have written that it is the old who teach the young. No longer. The young have broken away; they teach themselves and are creating a world of their own.

V. THE DUALITY OF MORALS

One of the foundation blocks of human behavior is morals. Our moral code is deeply ingrained by education in our early years. What are morals? Do morals correspond to intrinsic, absolute truths which exist and endure in themselves regardless of circumstances? In the writer's opinion, morals are but sets of rules which make living together possible, and depend on circumstances. In our western world, for instance, bigamy is one of the gravest of moral offenses, one that is severely punished by law; yet a few thousand miles to the east one might not be admitted to the country club without having a number of wives.

The main moral imperatives have been summed up admirably in the Ten Commandments of Moses. The main commandment is "thou shalt not kill." But nature is built on killing! Try to teach a tiger not to kill and it would, if it could, laugh at you. Teach a mouse not to steal (according to another commandment), and it would starve to death. Without killing and stealing nature would collapse.

Killing and stealing are part of the law of the jungle, but a human society obviously cannot be built on them. "Thou shalt not kill" (or else you will be killed too); "Thou shalt not steal" (or else you will be stolen from too, and will get nowhere). Both commandments are deeply ingrained in us in early youth; as a result most of us could not kill or steal without some inhibition.

The two above-mentioned commandments are the foundation of our social structure. The conflict created by these derived from the fact that man is a predatory animal, differing from other predators in that he also preys on his own kind. He likes to have peace at home, but likes to go out on predatory expeditions and dominate others. So he needs two moral codes, an individual and private one for use at home, and a collective and public one to be used on his expeditions. The two are diametrically opposite. What is shame in the one is glory in the other, and *vice versa*. While we may be punished for carrying an unlicensed gun at home, the state itself, which punishes us, invests a great part of our earnings in the creation of the foulest devices of killing, such as poison gases, napalm, and asphyxiating and defoliating agents. At present a considerable part of our youth is herded into prison for long terms of hard labor, or is forced to live in exile, for refusing to take part in the killing of poor and under-developed people ten thousand miles away who have never attacked their country. Senescent judges show how patriotic they are by passing out hard sentences for tearing up a draft card or following one's conscience according to the principles established by our country at the Nüremberg trials. Thousands of our young people languish in jail who have not preferred exile. Refusing to kill and tearing up a draft card brings the same punishment as armed robbery. Long years of hard labor in prison is legal murder, inflicting irreparable damage on the mind. But who will judge the judges, and who will govern the governments? I agree with Bernard Shaw* that every judge should go to prison for some time to see what they inflict on the people they send there.

This double moral code is generally accepted and is applied by governments in matters of foreign policy. Where

----

* G. Bernard Shaw, **The Crime of Imprisonment** (New York: Philosophical Library), 1952.

the difficulty comes in is that modern science has abolished time and distance as a factor in separating nations; on our shrunken globe today there is room for one group only, the family of man. There can be, thus, one moral code only. "My country right or wrong" worked only while nations were separated by distance. Since distance has disappeared and man is for all practical purposes bound into one group, we will have to call murder "murder," regardless of the color, uniform or passport of the murdered. Likewise, we will have to punish murder accordingly, whether single murders or mass murders, not excusing it even if politics are called in as the motive.

It is probably this dual code of morals which underlies the break in the career of many leading politicians who begin their political efforts with a desire to improve the lot of their fellow men. Once they reach the top they tend to exchange their individual code of morals for the collective one; they begin to serve abstract ideas, which have little to do with their people's well being, and they make war. The dividing line between a nation's glory and power and its leaders' own is not a sharp one. Collective human suffering easily becomes an abstraction too. I myself am touched whenever I see suffering, and death upsets me. I often go a block round to avoid a funeral, but the death of 100,000 people makes no impression on me. I just smile, being unable to multiply death or suffering by 100,000. This, to me, is a number, an abstraction. One death is a tragedy, 100,000 deaths are statistics. So it must also seem to men in high offices. From high up a human life must look very small, a notion that moved Walt Whitman to sing about the arrogance and audacity of elected government officials.

Unfortunately, this collective code of morals is not the exclusive property of those on top. We all share it as soon as, in some way, we participate in government. This is the case, for instance, when we go to the polls to elect hawks and vote the endless billions for war and the formidable machines of killing and destruction, and then go to church and ask for God's blessing.

VI. THE BIOLOGY OF ARMIES

One of the main biological features of an army, as that of a cancer cell, is that it has to grow and grow, even if growth is not needed; an army, with its closely knit organization and structure, acts like an individual, craving more wealth and power. There are two more reasons which make growth inevitable. One is that any army always creates a counter-army, and it has to keep ahead of it. In the end both armies strive to keep ahead of one another and go into an open-ended spiral of growth. The other reason for the growth of an army is that it has to be kept occupied, so it creates incidents, pulling its country into wars and doubtful ventures. It is this kind of activity that will lead to the scuttling of the United Nations, the only hope of mankind.

What we mean by "army" cannot be sharply defined. The army forms a single organism with the armaments industry and the government. The "military-industrial complex" is one inseparable unit which lives on the life blood of the nation which supports it, channeling the fruits of that nation's labor from more productive endeavors into its own unproductive adventures.

This growth is dangerous because if the military-industrial complex exceeds a critical mass it becomes the master of the civil authority instead of being its servant; it dictates the nation's foreign policy and the distribution of its resources, swallowing up great amounts of the national income; and it subverts all higher endeavor, causing art, science, and humanitarian institutions to wither away. Our military-industrial complex passed this danger mark years ago and its further growth has become autocatalytic, speeding up itself. The bigger and more powerful it becomes, the faster it grows. These military-industrial institutions modestly call themselves "defense departments" and claim that their sole object is to defend the unarmed citizen. In this country our Defense Department takes our youth ten thousand miles away from home to slaughter, and be slaughtered by, a people who have never attacked this country. It sees to it that anyone who objects to going out and killing should be brought to his senses by being put in prison. Most of us still have the rudiments of decency left in us and we don't like to kill. The military has a simple cure: it jockeys the reluctant individual into a position in which he has a choice only between killing or being killed, with only his reflexes to guide him. So we kill; sometimes we even overdo it. According to more conservative estimates it has cost the United States $50,000 to kill each Vietcong. For the same money he could have been sent on a luxury cruise ten times around the world.

Armies are a curse of mankind, a threat to peace, a threat to our very existence, a blot on the face of human culture and intelligence. The greater an army, the greater threat to peace it is. We maintain armies to solve the problems that arise between nations. But problems are like equations—they cannot be solved by bombs, not even by atomic bombs. It is a shame that we cannot create more intelligent means for their solution, and that we push aside the one we have—the United Nations.

I am not challenging the good faith of our army. It is created to defend us and it wants to do the job well, with the newest and best and biggest instruments. It cannot help it that military thinking is, by necessity, callous, narrow and shortsighted, having but one answer to problems— killing. The failure is with civilian governments the world over which use their armies for the solution of their problems, giving free rein to them. Armies, by definition, are the instruments of organized killing. Who will defend us against defense departments?

VII. THE DUAL WORLD STRUCTURE

To the superficial observer it must look as if the military world structure consists, essentially, of two great armies, the armies of the two superpowers, the United States and the Soviet Union, pitched against one another, holding one another in balance. The truth is different. The truth is that these two great armies are the sweetest of allies, for without the Soviet army we would not need ours, and *vice versa,* and the fruits of the labors of the citizens of these two countries would not go to their defense departments. The two great armies work in concert, promising fear and hatred toward each other to prevent peace from breaking out. They both are very brave and are afraid of only one thing, peace, which would make them superfluous. Thus they fight peace with all their might. They dominate not only their governments, but with their holds on the press also dominate and direct the minds of the people of their nations, patriotism with militarism, making it tantamount to treason to question any military expenditure, and equating patriotism with the act of voting huge sums of money for them. What we really need are defense departments against these defense departments, which use part of the taxpayer's money to make the propaganda needed to get more money out of him.

The way this whole system works can be illustrated by quoting from the October 27, 1969, issue of the New York *Times.* "Some Soviet military leaders, notably Marshal Nikolai I. Krylov, commander of the strategic missile forces, have said the United States is preparing for a surprise attack against the Soviet Union and the best defense is overwhelming strategic superiority." Briefly: the U.S.A. is out for a first strike, and to prevent it the Soviet people must have a bigger army and arsenal. These words of Marshal Krylov sound very familiar to us! They are, almost verbatim, what we hear daily from our own leaders, who talk about the Soviet strike capability, frightening the people on this side while Marshal Krylov does it on his. No attention is given to reports of the civilian information agencies, which find no indication that preparation for a first strike is taking place on the other side.

Of course, Anti-Ballistic Missile (ABM) systems are purely defensive, but an improvement in defense disturbs the balance of power and forces the other side to strengthen its arsenal. Thus it goes on and on, the money flowing into the coffers of the military-industrial complexes. It must be evident to everybody by now that there is no technical solution to the armament problem. If one side takes a step to add to its arsenal, the other side must take a step too, and so it goes on *ad nauseam, ad infinitum,* to the final catastrophe. Even a strong disturbance of the present balance would make no difference. Both sides have plenty of nuclear power to discourage anybody from starting an atomic war. Both sides could kill everybody three times over at least. A strong imbalance would mean that we (or the Soviets) could overkill only twice, which is still plenty.

The greatest danger, however, is not the big powers themselves. The greatest danger is that their lawlessness encourages everybody else to arm senselessly. Nobody can regard himself as a great power and expect to be talked to respectfully until he has joined the atomic club. So it is not surprising that the club, which originally had only two members, has four now; and the number of members can be expected to rise. If any of the governments having a nuclear capability gets into the hands of extremists, it may push us all over the brink.

But there is still one more danger lurking below the horizon. The Pentagon is engaged in a flourishing arms trade, selling weapons all over the world, inciting under-developed nations to join the arms race and spend their money on guns before they can feed their own children. The Pentagon is not the only one; the Soviet military authorities do the same, and even the supposedly defense-less Saigon government is engaged in a most flourishing arms trade. It looks as if a world cartel of armies is con-spiring to transform our whole world into a huge garrison. It could happen, and with it we would have all the drab-ness, callousness and moral stagnation of military life.

We fought World War II to eliminate militarism. We eliminated Japanese militarism, and now we labor to re-establish it; we press Japan to increase its military budget in exchange for Okinawa and economic concessions. The miraculous recovery of Japan and Germany shows what economics can do if released from the burden of arma-ments, while our own inflation underlines the opposite. Japan, with no army and a refutation of war as an instru-ment of policy, is today the safest country of all, as far as there is safety for any nation. The only thing that en-dangers it is American bases on its territory, which turn Japan into a target.

When I was a young man there was a great boxing champion who, when asked if he didn't feel safe because of his skills, answered that "a polite tongue gives more safety than a strong fist." So does good will.

VIII. ON GOVERNMENTS

Much of what I have said about armies holds true for governments as well. This is not only because governments are more and more dominated by their armies, but also because governments need enemies to frighten their people with, frightened people being more easy to lead.

It was two hundred years ago that the Babeuf decreed governments to be "conspiracies of the few against the many." His government readily supplied the proof for the correctness of his statement by having him decapitated. And of course conspiracies between governments against other governments are not unheard of. The American effort in South Vietnam is clearly illustrative of this.

Occasionally, these days, we see what appear to be moves towards peace, such as the Paris peace talks, the renunciation of bacteriological warfare, or the conference on the limitation of armaments. It is questionable, however, how much these actions are genuine peace moves and not merely sand thrown into our people's eyes to make basically hawkish policies more palatable. President Nixon himself has discounted the Paris peace talks. His unilateral renunciation of bacteriological warfare was a gesture in the right direction, but its value is reduced by the fact that it was made after it had been shown that such warfare is equally dangerous for both sides; also the renunciation was made immediately after the My Lai mass slaughter had become known and something had to be done to polish our badly tarnished reputation. Budgetary measures indicate that research on bacteriological warfare will go unabated, only its label will be changed to

"defense research." As to the conferences on the limitation
of armaments, these could safely be initiated without the
danger of drastic consequences. It is not possible to put
the cart before the horse: reduce armaments and then
make friends. One has to make friends first; once that is
done arms will become obsolete by themselves, there
being no use in arming against friends. The armament
race has no technical solution. The situation is similar to
the situation in Texas: it was suicide a century ago to
throw away one's gun; then the psychological situation
changed and guns disappeared by themselves. Today one is
safer in Texas without a gun. According to Lincoln, the
best and cheapest way to get rid of an enemy is to make
a friend of him. Why not make friends first with our
enemies? The American government carefully avoids any
step in this direction, which would make armaments really
superfluous.

The conferences on limitation or reduction of armaments
have their own humor: we prepare for them by speeding
up our arms buildup and research (MIRV, ABM) so as
to "deal from a position of strength." Disarmament means
less arms, not more.

Why governments do not represent the people which
elected them is a rather intriguing question. There may
be many reasons for this. One reason may be that to be
elected, one has to be a good politician, versed in all the
tricks of the trade, while to be a good leader one has to
be a good statesman. Politicians think of the next election,
while statesmen think of the next generation. People elect
the best politicians and then are astonished when they
discover they have gotten poor statesmen.

But there must be deeper reasons to explain the turn
politicians take when they are elected to office. As a candidate, President Johnson faithfully promised not to send

40

our boys ten thousand miles away to do the fighting Asian boys should have been doing for themselves. He set out a plan for a "great society" which could have made him into one of our greatest Presidents. But once he was elected, he made war, squandering the means for the building of his great society. President Nixon, as a candidate, promised that his first trip abroad would take him to Moscow to make friends. As President, his first trip took him to Bucharest, to "put pepper under the nose of the Soviet," as the Hungarian saying goes.

It would be unfair to state that all the promises of candidates are plain lies. There seems to be an abrupt change of course in a politician when he gets elected to office. One reason for the change may be the unwitting exchange of his individual moral code for the collective one. Also, from the top things must look very different, and a single human life must seem very small. Until the candidate becomes a representative of the people, he sees only individual suffering. Once on top suffering becomes to him statistics. Perhaps this is why Mao Tse Tung can quietly calculate that if in an atomic war 100,000,000 people are killed on either side, half of the dead would be Americans, while most of the Chinese would be still alive, making it a good bargain.

President Kennedy thought that the Cuban Missile Crisis was a 25 per cent gamble. I do not want anyone to gamble with the life of my children and grandchildren and the whole of mankind with 25 per cent probability. There are other less risky solutions.

There is a growing dissatisfaction with this state of affairs and there is a growing willingness in people, if not desire, to stretch out their hands toward the "enemy" in friendship, over the head of their government. The November Moratorium in the U.S. may have been the first massive expression of this trend. More than a quarter of a million young people marched on Washington. This march may turn out to have been the most momentous event in our history. The sight of Washington on that day reflected a queer spectacle: the blocks surrounding the White House were closed off and deserted, the White House itself was protected by a circle of busses, lined up bumper to bumper, reminiscent of the wagon trains of our pioneer days preparing for an Indian attack. A few hundred yards further out were massed the quarter of a million youngsters, elated, with the idea of peace in their minds. What will happen when they grow up to be the government themselves? They may replace the slogan of the last century, "proletarians unite," with "people of the world unite."

IX. ON INFORMATION

It is said, and not without reason, that "he who understands the situation is not well informed." This statement could also be turned around: "he who is well informed cannot understand the situation." I shudder to think of the mass of petty information which must reach our army and government daily from all their various secret information agencies. It has to clog any brain which tries to take it in. I suspect that one of the aims of "information" is to keep the public out of the business of government, telling it that it is not informed. This information is like the medical recipes we used to write in Hungary in my student days. We had no end of entirely useless drugs and we prescribed them in the most complex recipes, all in Latin, so that our patients would not understand them, and would thus be kept in the dark about our business, which we tried to keep a mystery.

Really important information is accessible to everybody, except to the army and the government, which are preoccupied with their secret reports. A few years ago I spent a few days in France, participating in a scientific conference. A middle-aged French actress was assigned to me as hostess. It was inevitable that our conversation should turn to Vietnam. "You can do nothing with the Vietcong," she said. "They fight and fight and fight. They die, but don't give up." This was real information. Her knowledge was gained as a result of the protracted war of the French and the Vietnamese, and forty million Frenchmen knew it. Only our Pentagon and White House did not get this information, and expected a final victory at every step of escalation.

Another piece of information relates to power. Between the two world wars, at the heyday of Colonialism, force reigned supreme. It had a suggestive power, and it was natural for the weaker to lie down before the stronger. Then came Gandhi, chasing out of his country, almost single-handed, the greatest military power on earth. He taught the world that there are higher things than force, higher even than life itself; he proved that force had lost its suggestive power. This, again, is a piece of information which did not reach the Pentagon or the government: we cannot win in Vietnam because the people are willing to die faster than we can kill them.

Here are a few more important pieces of information: the road to peace is not supported by bombs, napalm, defoliants and dead bodies, but by good will and human understanding. The great "silent majority," or, rather, "silenced majority," of mankind is longing for peace and will accept as its leader the one who is best equipped to lead it to peace; it will not accept as its leader merely the strongest or most bellicose. And here is another piece of most important political information: the best policy, in the long run, is that of simple honesty and sincerity. Trickery, double dealing and double talk are, in the long run, self-defeating.

There is one great danger about information which became clear to me when, during the short democratic period of Hungary, I sat once on the dais with some government officials during the celebration of a national holiday. My neighbor was Rákosi, the later communist dictator who was at that time preparing to take over. Shortly before, he had been to the United States to spy on our Prime Minister, who had been invited there and whom he accompanied. He was shown all that is really worth seeing. During the President's speech, I felt a tap on my shoulder. It was Rákosi. He wanted to say something to me, so I bent back to listen to him. "I was in America," he whispered, "and I have seen a lean horse and a man in torn trousers. Not everything is as good there as they say." It was this that he wanted to tell me, and I quote him verbatim. I was dumbfounded. Rákosi, in a way, was an honest man. He was a criminal by all standards, but in a way he also was an honest man who would not lie when lying was not needed. I believed him, that all he saw was a lean horse and a man in torn trousers. What I am driving at is that we see only what we want to see, and hear only what we want to hear; and I expect that the secret agents who pick up information for our army or government see only what their bosses like to hear.

Another kind of information which none of the governments of the big powers seems to have is that high-energy radiation damages DNA,* the genetic material, the most sacred treasure of mankind. The information is not to be found in secret reports, but every child knows it. It has to change the whole outlook on war and the future of mankind.

DNA, the genetic material, is the most wonderful thing in the world, guarded by nature in the most careful manner. Mankind went through epidemics, famine, and all sorts of trials, yet nature kept this material intact, because all life depends on it. Today is the first time in history that man has found a means to damage it. High energy radiation does so. This makes an enormous difference in the nature of atomic war. In earlier wars the DNA was not directly threatened and mankind could carry on life. There may also be survivors after an atomic war, but those survivors will be unable to produce a healthy progeny. Their progeny will be beset by abnormalities, monstrosities and

_____
* Deoxyribonucleic Acid.

diseases which will make life not worth living, and there will be no way back.

An American senator said recently that the important thing is that if we have to start with Adam and Eve again, they should be Americans. They will not be Americans, only utterly miserable, destitute members of a dying race. This information evidently has not reached the halls of government yet. So our governments still add daily to their terrific arsenals which if they remain unused are thus a waste, and if used, will destroy us all.

Man is a strange creature. If I were to invent a new safety belt for cars that would increase the driver's chance for survival by a hundredth of a per cent, everybody would run to buy it. Yet we do not protest when the chances of our survival rapidly decrease to 50 per cent with the further stuffing of our atomic arsenals. Our lives and those of our children are in hands which we rightfully neither know nor trust.

In order to get a clear view of the world, it is not the enormous bulk of official secret information that has to be studied. There is no clear view from under those heaps of papers. To get a clear view Adlai Stevenson sat for hours on a hill, looking out on the surrounding country, his America, taking in the wide and distant horizons. Today we have an even better means of seeing the world clearly. All we have to do is gaze and reflect upon the pictures of our planet that were taken by our moon voyagers. The real information is there, open to everybody. It tells us that we are all occupants of a tiny globe and that we should be constantly at war is absurd.

X. LIFE vs. DEATH

The primary aim of science is to find truth, new truth. This search is the more successful the more it is directed towards finding truth for its own sake, regardless of its possible practical use or application. All the same, new truth and knowledge always elevate human life and most usually find practical application. As a rule, the more fundamental and abstruse a new truth, the greater and more important will be its practical possibilities. In fact everything we have, including life itself, we owe to science, to research. If everything given to us by research were to be taken away, civilization would collapse and we would stand naked, searching for caves again.

Even pure truth, which has no application whatsoever, elevates life. From a practical point of view it is quite irrelevant whether the earth turns around the sun or the sun turns around the earth. All the same, Galileo and Copernicus, by their discoveries, lifted human existence to a higher level. The same may one day be said for our moonshots. Science is life-oriented.

Contrary to this, armies and armaments are death-oriented. Armies are instruments of organized manslaughter, in whatever way they may be used, for defense or aggression. All its tools are tools of death, be they guns, bombs, napalm, tanks, missiles, bombers, or gases. They all are instruments of killing. So the military is death-centered, and a society dominated by the military is a death-centered society, as pointed out by George Wald in his famous Moratorium Speech.

Tools can be used for both construction and destruction,

for elevating life and for destroying it. The more power-ful a tool, the more it can elevate life, but the more extensive the killing and destruction it can achieve. The achieve-ments of science intended to elevate life have been con-verted by the military into instruments for destroying it. We biologists have achieved a wonderful knowledge of how our nerves work; the military used our knowledge to make nerve gases. We have achieved a wonderful know-ledge about the nature of disease, especially infectious disease; the military has used this knowledge to perfect bacteriological warfare devices. We have achieved a wonderful knowledge of plant life; the military made defoliants with this knowledge. We disclosed the hidden energies of atoms, to elevate life and eliminate toil; the military made atomic and hydrogen bombs with that knowledge, with which it can wipe out mankind. This makes the military-centered society a death-centered society marching towards doom. We all may end up one day like the 6300 sheep found dead in Skull Valley, Utah, one morning after a military plane spraying nerve gas slightly miscalculated the wind and altitude. The error was so slight that the Army had difficulty in finding it and at first denied any responsibility. What will happen in case of a major accident, or in case of a war in which op-posing armies deliberately open their deadly taps or fire their poison-laden missiles?

While irreparable damage is inflicted on science in the United States for the sake of an economy of 100 million dollars, 80 billion dollars are spent yearly on military concerns. It was just the day before I wrote these lines that Congress voted 20 billion dollars for "extras" for the Army. Twenty billion is two hundred times a hundred million. The relationship between military spending and spending for the betterment of society was nicely illus-

48

trated not long ago by the New York *Times,* which had two columns side by side on its front page. One column reported a minor scientific discovery: the development of a vaccine against German measles, which was expected to save 30,000 lives yearly in this country alone. The adjoining column brought out the weekly casualty statistics in Vietnam; saying it revealed that the number of G.I.s killed had reached 40,000. The discovery of the measles vaccine may have cost in the neighborhood of a hundred thousand dollars, while to have those 40,000 G.I.s killed had cost a hundred billion.

When I discovered Ascorbic Acid (Vitamin C), I felt proud to have made a contribution to science which could, in no way, contribute to killing. My pride was short-lived, however. One day, while visiting a factory, I noticed a collection of large jars and was told that they contained crude preparations of Ascorbic Acid. These were placed in German submarines and enabled them to stay at sea for months on their death-dealing missions without the crew breaking down with scurvy.

We constantly cut and cut scientific and cultural appropriations to enlarge our already swollen military budget. A society which is death-oriented is difficult to save. In an atomic war, the only fortunate ones will be those who die with the first bang.

The only hope we have is that the government-supported world military conspiracy will provoke a backlash, that the peoples of the world will rise in a revolution against it and stretch out their hands towards one another. We must shake off and chase out all armies and all the hawks who support them, chase out the maniacs of destructive gadgetry, before we can build a better life by using the wonderful tools and possibilities offered by modern science.

XI. ON VIOLENCE

The inclination of many of our people to violence is one of the main symptoms of the present world crisis. Violence is a terrible nuisance and we do not know how to deal with it. When President Johnson commissioned a conference to find out the reasons for the sharp upswing in violence in the U.S., it was typical of the hypocrisy of our age that none of the members of that conference was able to see that the violence of our times may have something to do with Vietnam, where we have been busily occupied with the killing and maiming of half a million people. Though Vietnam is the main focus of our putrefaction, I do not think it is the only source of violence. We have gone through two world wars and have watched Hitler and Stalin each exterminate countless millions of innocent people. Violence seems to be a symptom of the imbalance of our age; we have only added one more crime to the major organized crimes of our age by the Vietnam war.

There are two kinds of violence: active and passive. Active violence is the violence of people who break windows or skulls. Passive violence is the violence of those who would yield to nothing but active violence. The two kinds of violence are most intimately connected: it is passive violence which begets active violence, leaving no other way open.

President Johnson never broke a skull or a window, so according to the usual standards he was not violent. But if we extend the idea of violence to passive violence, then he must be classed with the most violent figures of political history. When 30,000 people marched on Washington to bring their wishes before him, he did not even ask what they had on their minds and received none of their leaders. For 30,000 people to leave their homes and march to Washington in slush and rain would indicate some very strong feelings of outrage on the part of a significant portion of our citizenry. The first duty of any president is to listen to his people. The 30,000 marchers were obviously only a very small fraction of the people who shared their feelings. The only acknowledgment President Johnson gave them was his insistence that they march in order and avoid violence. For so long as they respected this edict, they could march until doomsday. He began to pay attention to them only when they tried to break into the Pentagon, that is, resorted to active violence. It was Johnson's passive violence which provoked their active violence. President Nixon went even further, declaring well in advance that the marchers would not make the slightest difference to him. Then he barricaded himself inside the White House.

Violence on university campuses is of special significance. The way to prevent it is to avoid passive violence, which means that our authorities must listen to the protests of those they supervise with sympathy and human understanding. Students' demands may often seem excessive and ill thought-out; nevertheless, they have very deep roots. Our youth is our only hope for a better future. Why is it that university faculties can consent to reforms only after active violence has taken place? This shows our youth that active violence is the only way to get action. Police or disciplinary measures, if applied at all, should be called on only after all other avenues of understanding have been exhausted. This should be the case for any group disturbance based on protests against existing conditions.

XII. GERONTOCRACY

Konrad Lorenz, the great student of animal behavior, hatched goose eggs at the foot of a chair, and the goslings recognized the chair as their mother for the rest of their lives. When put under the chair a few hours after hatching, there was no such reaction. The point of this experiment was to show that things can be imprinted into the brain at an early age only; the brain freezes up later and is no longer malleable. In dogs this freezing up occurs around the sixth month. If you want a dog of the wolf family to recognize you as his master, you must train him before he is six months old. This freezing up, in man, seems to occur around the fourth decade, after which the brain is increasingly unable to assimilate new ideas.

Max Planck, one of the greatest scientists of human history, the father of the quantum, wrote in his autobiography that it is impossible to convince people of anything new. All that one can do is to give them time to die. The young generation will then embrace the new truths.

My beloved mother was an enlightened agnostic who just smiled when people talked about religion; but if any of her sons were in trouble, she hastened to church to bribe St. Peter with a dime so that he would lobby for her. The impressions of her early childhood were undeletable; later impressions constituted but a thin layer which peeled off easily.

I experienced this myself. I was in my fourth decade when I started to work in quantum mechanics and tried to understand the atom. It was too late. I could pick up ideas with my brain, but they never got into my blood and I found that I had to avoid discussions on atomic physics with high school students. They have the atom in their blood as well as in their brain.

Although modern science made its appearance at the turn of our century, it was not before the fifth decade, after Hiroshima, that it became the life blood of our age. People who reached their fourth decade before Hiroshima can never grasp what atomic power means. They may learn about it and may even know by heart the radius in which bombs of various mega-tonnage will destroy life, but the idea of atomic power never got into their blood. Their knowledge does not become part of their being, it is book-knowledge only. It is a different situation altogether with children who, when growing up, have to pass air raid shelters every day when going to their classrooms and know that "air raid shelter" means that one day they may be herded into it and find, on emerging, that their world has disappeared. I am sure that many of our present political leaders know much data about hydrogen bombs, but their blood still flows with the old world, with old-world ideas and perceptions. An atomic or hydrogen bomb remains for them only a bigger and better bomb.

Our present world is a gerontocracy, dominated by people whose brains froze up before the atomic age. They do things which may have been right to do before this age, but have no relevance in the new order of things.

I carefully watched on my TV screen both the Democratic and Republican conventions of 1968. I was struck by three things. First, I saw no young people. Fifty-five per cent of the people, thus the majority, of the world today are under 30. In South America, 50 per cent are under 21. At these conventions I saw no one under 30 and hardly anyone under 40. The majority was clearly not represented, and it is obvious that it is excluded from our daily political life. Indeed, people under 21 do not even have a vote.

We are a gerontocracy. Gerontocracy is a good system at times when changes are slow and the main problem is the preservation of values, but gerontocracy becomes highly dangerous in periods of rapid change, such as the present period represents, when man's existence depends on his ability to adjust to and create a new world. As a schoolboy I often amused myself during boring classroom sessions by bringing together, in my mind, the leading figures of different ages. I found in those days that Julius Caesar and Napoleon would understand each other perfectly and discuss all their political and military affairs without difficulty. History was stagnant for two thousand years. Today they would both be complete strangers.

The second point I noticed about our two national conventions in the summer of 1968 was that there was no discussion about principles of government or any of the great problems of our day. It was purely a question of power, who would be in and who would be out. Our political parties, which were created to establish and sustain the principles of government, have become nothing but the instruments of personal ambition. It does not surprise me, then, that the vital problems of mankind, like the ABM, which added a new turn to the armaments spiral, are decided by petty horse-trading tactics—selling votes to the President for favors or advantages; promising votes in exchange for the dismissal of an able man from a governmental agency; trading votes for a delay in school desegregation in the South. *Parva sapientia regitur mundus.* It is poor wisdom which rules the world.

The third thing I observed was that there were no women at the conventions. Where *were* the women? Women have more common sense than men, and they also have votes which are just as good as men's. Moreover, they are the ones who produce the fodder for the guns. Why do our women, longing for "liberation," not get together and do something about all those hawks who send their boys to die. The Vietnam war has already claimed more than 40,000 young lives (not counting the Vietnamese) and a quarter of a million wounded or maimed. Everybody has a mother, so these figures mean an equal number of frustrated parents, sweethearts or wives, and children who will have to grow up without fathers.

Gerontocracy, it seems, searches for the future in the past.

XIII. SEX

Sex and hunger are the strongest appetites in man and therefore evoke the strongest feelings in him. While hunger has often led to war and revolution, sex has not. It has no political import; it is important only ecologically. All the same, it is the strongest driving force in life; without it, of course, life would cease.

In sex, the sublime and the vulgar are separated only by a hair's breadth. Christian religions could never find a consistent attitude towards it, making it a sin before marriage and giving it their sanction after marriage. Overlaying it all was the residual feeling that anything having to do with fornication was evil—this is the legacy that has been passed down to us by religion. Today, our young people finally seem to be shaking off this anti-natural legacy; perhaps as a result the world will one day soon see a more sensible attitude towards sex.

At the age of 76 I do not feel I have the qualifications to talk on sex today or pose as an authority on the subject. What I can talk about is sexual morality as it was at the time when I was young. Ideas about sex in those days were in an awful muddle. Sexual intercourse, outside of marriage, was a sin. Chastity was a virtue. But if the urge became too strong, all one (a male) needed to do was go to a brothel, which was an accepted social institution. In my class of society one was not supposed to talk to a "decent" girl, even in her parent's home, without the presence of a chaperone, usually some elderly lady. Girls were considered the more attractive the less they knew about sex.

57

All these customs had the pretense of morality, and caused no end of suffering, creating unbalanced minds and psychopathic aberrations.

Then came penicillin and the "pill," and the entire moral code dealing with sex began to be blown away. The fear of pregnancy and venereal diseases would appear to have been at the heart of our old morality, that morals are but prescriptions which make a society possible. A society loaded with venereal disease and illegitimate children would not be a stable one. All the same, the sudden change in our sexual morals shows how shallow some of our moral convictions are. It also shows that history is no longer made in the capitals of the world, but in the laboratories where such things as the pill and penicillin were discovered. The liberation of man from his painful sexual restrictions is one of the major turning points in the history of human life.

In my view the most wonderful achievement of our youth—a sign of their great moral courage—has been their ability to restore the sexual drive, the strongest of human feelings, to its purity and dignity. They have made human life much richer and cleaner, and have made me wish I had been born fifty or sixty years later.

It would be interesting to go through the whole gamut of our moral convictions in order to see whether they possess the same shallow foundation as our convictions about sexual morals did, and to see whether they cannot be replaced by something better, cleaner and richer. Whatever hope exists for the world resides in the aspirations and abilities of our youth. If our youth were able to create a new, more wholesome sexual code, if they were able to reject traditional and ingrained pretense and hypocrisy in sexual matters for frankness, honesty and dignity, there is no reason why they cannot be the creators of a new morality in other areas as well—areas on which our very existence depends. They are trying—trying to replace narrow nationalism with human solidarity, trying to replace war with peace. If I were to pray at all I would pray that they succeed, that they never give up, in spite of the punishment they must endure.

XIV. THE GENERATION GAP

This is one of the most important problems of our time. It is a very complex one, too. Its complexity derives from the interaction of various independent factors, political and biological, which we will have to consider separately.

The generation gap is part of the general tumult of our time and so it has to share the common foundation of all our troubles—the emergence of modern science. Up until recent years the change in human outlook was very slow. There was a lazy continuity to history, and there was always plenty of time for adaptation. This situation changed, to some extent, with the emergence of classical science which, with a latency period of a couple of centuries, led to the Industrial Revolution of the 19th century, bringing upheavals and serious changes into man's life. Modern science emerged at the turn of this century and needed only half a century to start changing human life, changing it at an increasingly rapid rate. This situation was dramatized by World War II, which held up progress for several years and yet was the trigger of rapid post-war change by virtue of the fact that it brought modern science into our lives with a bang. I mean "bang" literally, because it was the explosion of the first atomic bomb at Hiroshima that told man that life would never be again as it was before. That fateful event already overshadowed the later differences between the military and scientific outlook with respect to the capabilities of modern science. Most of the scientists who worked on the development of the bomb demanded that it should not be used until all other possibilities were exhausted. They insisted that a demonstration be given to the Japanese government of its power before using it against Japan. The military point of view prevailed, however, and the first bomb was dropped, killing 100,000 people and starting up a nuclear armaments race that, since that time, has consumed most of the fruits of human labor and brought mankind to the brink of extinction. The post-War world is patently different from the pre-War world, and the people who grew up before the war have never been able to understand those who grew up after the war. The Second World War represents a discontinuity in human history and tradition. This is what underlies the Generation Gap.

This gap is further widened by biological factors. The most important factor is the freezing up process of the brain, to which I alluded previously. The people who hold the power and direct the policy in this country are practically all over 40, pre-War people with the old world and its values frozen into their minds. They all witnessed Munich, where the green light was given to Hitler, and they are trying to do now what should have been done then. They try to run today's world according to the rules and experiences of a world that no longer exists.

Another biological factor that contributes to the widening of the gap is the fact that man, during his individual development, has to go through different stages. A newborn baby is an entirely different individual than a grown-up. It has its own particular needs and rights, and these are different than those of persons in other general age groups. The same holds true for the other age groups—they are all different from one another; they have different psychologies, different needs and requirements. A child is a child, a young man or woman is a young man or woman, a grown-up is a grown-up, and an old man is an old man. Correspondingly, all these have characteristics that separate them from each other.

One of the strongest traits of man is his inclination to dominate. The stronger dominates the weaker. For a long time men dominated women, and grown-ups dominated children. "Children should be seen, but not heard" was the Victorian educational slogan. Today's youth have broken through these limitations and demand their own rights.

Our children come into this world with "clean and empty minds." What they learn of the world as they grow up is markedly different from what the children of the pre-War world learned. Today's adults look at the world through the glasses of pre-War and pre-scientific values. They think that all the world needs is a little bit of patching up to make it work. The result is that we get deeper and deeper into trouble. The modern scientific revolution has made all human institutions age faster and faster; as a consequence we have a hypocritical world where everything serves a different purpose than it professes to serve. Our youth rejects this anachronism wholesale. For what do they see? Going to their classrooms they have to pass their school air raid shelters, which experience reminds them that one day their whole world may be wiped out. They are driven to the conclusion that there is no value in endeavor and that the only wise thing to do is to enjoy the moment, enjoy their growing bodies, enjoy what life can offer for so long as it lasts. They see the fruits of labor converted into monstrous instruments of death which can wipe out all life. They see "the energies and resources of nations diverted into ever larger ways of expunging and cheapening human life," they see "people preoccupied and swollen with meaningless satisfactions," they see "the concepts of human brotherhood and social justice held up as . . . animating ideals, but find that their own efforts to act on behalf of these concepts will put [them] in conflict

63

with that same society," they see "a world insanely fouled by pollution . . . the good earth being covered over with tar and cement, the streams and lakes being poisoned by detergents and chemical excesses, the oxygen in the air being depleted. [They know] that our astronauts were able to roam the heavens because no limitations were placed on human ingenuity, technological facilities and funds and [they are] unsatisfied with the argument that it is impossible to make life on earth a little less hellish."*

They find everything a lie. The great political parties are out for profit and power, the military for domination, fattening itself with their young bodies. They find the churches preaching love but raising no voice against the slaughter of an underdeveloped people; they also see the churches driving the world towards overpopulation for the sake of religious imperialism—resisting family planning, denying children the right to be received into this world wanted and welcomed with love and care. They see that religions are always on the side of power. And they see that while half of the children of the world go to sleep hungry, without the food they need to build sound minds and bodies, we spend hundreds of billions to raise our stack of nuclear bombs and missiles higher and higher. They see that most of their political leaders are really mindful only of their re-election, of keeping their power, feeding the people with arguments which should be rejected by the simplest logic, refuting the great ideals on which our country was built.

The only thing youth can do in the face of all this is to reject the old-world values wholesale. And yet what a sense of frustration they must feel when they see their

---

* Quoted from Norman Cousins in **The Saturday Review,** November 8, 1969.

efforts to improve things get shot down summarily by the adult population, which is content to see the world continue in its schizophrenic balance—after all, today's over-40 adults will most likely not be here when the Armageddon they have been creating finally explodes on today's youth.

I wish that our youth were more conscious of their power because, whether they know it or not, they do possess a considerable amount of it. Our youth must win the battle they have engaged themselves in because the old ones have to die and youth will have to take their place. I wish that youth would start preparing themselves for building a better world right now. I wish that representatives of the youth of different countries would get together and hammer out the constitution of the world they want to build. The future is theirs, but they must start building it right now by preparing themselves. They must not only prepare by planning for the future but also by taking immediate action, hunting out of their governments the merchants of death and destruction. They can do it.

It is most important to start right away because there is the danger that the present rulers, when they feel their grip being loosened, may try to strengthen their hold by another World War or by playing into the hands of fascism. It is known that Hitler, when he recognized his impending defeat, planned to destroy the whole German nation, which seems to be a normal reaction for a madman. Are our present leaders any less mad?

Efforts have been made to close the generation gap. It can be bridged to some extent, and open conflict can be avoided; but it cannot be closed. It should not be closed. This gap is mankind's only hope for the future. The present world is in a groove which is propelling it to doomsday. In order to survive we must make a sharp turn; but so far we have not been able to turn, the groove is too deep and narrow. A new beginning has to be made, but only our youth can make it. The transition can be eased by giving youth more rights, since it is their world too. The present situation is nothing short of ridiculous. This country set out on its history with the cry, "no taxation without representation." The highest tax anyone can pay is to give up his life. Our youngsters between the ages of 18 and 21 can be taken by force of law to a faraway Asiatic country to die for a corrupt, autocratic police state without having a word to say about it. Old men show off their youth and vigor by their severity at draft boards, and old judges show how patriotic and vigorous they are by giving youthful resisters to this insanity hard sentences. Over forty thousand of our boys have been killed and a quarter of a million maimed and injured, while thousands are in exile or languishing in prisons for having refused to participate in something which is generally recognized as a disastrous mistake and is regarded by a considerable part of the world as one of the major political crimes of history.

Many of our youth today are being forced to give up their lives for mistakes made by their elders, yet are prohibited from obtaining a drink at a public place or voting for the leaders who send them to their doom. Similarly, heavy penalties are imposed on the relatively innocent drugs of our youth, while the more dangerous drugs of the old, such as alcohol and tobacco, go free. Of course, most of the drugs of our youth do have ill effects—certainly some of the social effects of marijuana are not desirable. However, there is only one medicine that will be effective against drug use: a livable world, a restoration of faith in life—its dignity, value and longevity. Police raids and jail terms are not the answer.

XV. A PRESIDENTIAL SPEECH

(Which Was Never Delivered)

"It is more generally recognized that the very existence of mankind is in danger. It is endangered from three different sides: war, pollution, and over-population. As a result of these dangers many of our most outstanding scientific thinkers count the probable future life span of our race in years or a very small number of decades. I hope they are wrong, but they may also be right, which makes it my duty to use all means at my disposal to avoid these dangers. Compared to these, all other problems are of no import.

"Of the three dangers, war is the most acute one, not only because it can put an end to our existence in one blow, but also because it taps the resources which we need to meet the other two.

"We must put an end to all war, and since we are currently engaged in a war of our own, our first duty is to put an end to this one. The problem is a difficult one. The main difficulty is not that we must try to save face. The main and most real difficulty is that we cannot leave this unfortunate country with decency without having created peace. Our real ally is not the South Vietnamese government but the South Vietnamese people, for whose fate we made ourselves responsible. Time is running out and fast action is needed. For this reason I have urgently asked the South Vietnamese government to release all its many thousands of political prisoners and discard police-state methods. I have asked the present Saigon government to extend its political base and form a more popular coalition which will then conduct a fresh election in which the people can declare their will. I will use our armed forces there to insure the honesty of this election. Our great democracy cannot remain the ally of a government that uses suppression to gain its objectives. I hope the South Vietnamese government, for which we have made such enormous sacrifices, will heed my request. Should it refuse to do so, then we cannot remain its supporter and will have to withdraw, making this government responsible for its own fate. Should it accede, we will then be able to use our influence to prevent bloodshed and destruction, and help to build up what we helped to destroy.

"What we do henceforth in Vietnam will depend on the results of these elections. The people can tell us whether they want our presence there. The people have never been asked this question, although they have always been promised the right of self-determination.

"I have considered other possibilities. I have been advised to build up the South Vietnamese army to enable it to take over its own fight and enable us to pull out. This solution is not a good one. It is no solution at all; it means, rather, perpetuation of the war and would probably force us to leave hundreds of thousands of our young people in that country.

"In order to show the sincerity of my intentions, I have given amnesty to all draft resisters. I have ordered those in prison to be released, and I am inviting those in exile to come home to help us build the future together. I am responsible not only to you and to our young men, but to the whole world, to that great 'silenced' majority of mankind which wants peace and wants to see an end to war, bloodshed, and the senseless accumulation of the instruments of death. Our Constitution compels us to a decent respect for their opinion. This country has always had an alliance with life and not with death, and is responsible not only to the present generations but also to all future ones."

XVI. SCIENCE AND SOCIETY

As I pointed out at the outset of this volume, the quality and level of human life depend on man's relation to nature: how much he understands nature and can use its forces to his advantage. In earlier ages this relation depended on individual experience. In our age it is represented by science which, owing to its sophistication and complexity, is no more the common property of the people, but the property of the scientific specialists. Scientists have, in large measure, become the determinants of the quality of life in society. More history is made at present in laboratories than in national capitals. The last world war was not won by our armies but by our scientists. That Hitler did not win the war we can credit to the fact that he distrusted science. He had all the means necessary for the development of the atomic bomb before we did, and of course it would have assured his world supremacy. Had Napoleon listened to James Watts he could have conquered England. It is fortunate that tigers have no wings.

All the same, it is understandable when taxpayers and congressmen are peeved at being compelled to give money for something they do not understand, which to the outsider may look like nothing more than a bunch of men loafing around in white coats. It is important to understand what science is and how it works.

Some time ago I read an article by Warren Weaver, who counted the dynamos in his home and found 12 (if I remember the figure correctly). What nonsense, I thought! What do we need dynamos at home for? Then I set out to count the dynamos in my home and found

16. Without them my household and the quality of my daily life would collapse. This made me realize how much our daily life is dependent on science, how much we owe to science for everything we have.

We hear and talk much about organized and directed science, but most scientific progress in the past was due to single individuals who became fascinated by problems to which they then gave their whole minds. When Newton was asked how he made his discoveries, he answered, "by always thinking of them." This cannot be done on order. It cannot be done for profit either, because basic problems and their solutions do not usually lead to immediate application. After a while most basic discoveries do find application and thus contribute to the improvement of life. The lag period may be years or decades. It tends to shorten, however, as time goes on. The whole industry and technology on which our civilization is built have their origins in basic research. Even the steam engine is based on such research (the Carnot cycle of thermodynamics, discovered in 1824; it is this cycle which also underlies the development of the Diesel engine). The dynamo, too, belonged for a while to the order of useless scientific toys. The story has it that when the British Prime Minister, Gladstone, asked Faraday what use it had, the only answer Faraday could muster was that "it may be taxed some day."

Newton's friends were often worried about his mental state when they saw him sitting motionless all day long on a bench in Trinity Court, Cambridge; and the taxpayers may have understandably become indignant when they were asked to support such a "do-nothing." The lag-period between discovery and application makes it appear as if basic research were entirely useless. Today this makes research into a sitting duck for congressmen who seek to cut scientific budgets in trying to show their concern about the taxpayer's dollar. If all our industry (defense industry too) is based on the basic research of, say, ten years ago, we could cut out any further basic research altogether without any direct damage today; but ten years from now industry, even defense industry, would crumble. Damage would become evident only after it had become irreparable.

Where creativity is involved, the business experience does not work. Businessmen may argue on the basis of efficiency: that if one woman produces a child in nine months, nine women should be able to produce it in one month. Where creativity is concerned, businesslike arguments do not work. Paradoxically, basic research is the more useful the more it forgets about practical applications. Whenever young men come to me saying they want to go into research because they want to help to decrease human suffering, I advise them to go rather into charity. Research wants egotists, fascinated by "useless" problems, willing to sacrifice everything, including their lives, for a solution.

Of course, this does not mean that science cannot be or should not be applied to the practical problems of our day, but we must distinguish between basic and applied research, and give both their place.

The relationship of basic research to society is obscured by the fact that many discoveries of science often influence society in a very roundabout way. Newton laid the foundation of our knowledge of light; then came Huyghens, Maxwell, Herz and many others to establish the idea of electromagnetic waves. The result is that we have color television in our homes. That is the technological result. But what of the human result? Even the poorest person can own and watch a TV and see how rich and exciting life can be; in doing so, however, he will find it hard to understand why he has to live in squalor and misery if life is not such a vale of sorrows, as he has been told for centuries by the wealthy and powerful. So it is possible for the basic work on the theory of light eventually to lead to one of the greatest of revolutions—the "revolution of expectations," which may one day transform the face of human life.

Anything we have today, including our lives, we owe, in the last analysis, to basic scientific research. If the fruits of this research were taken away, civilization would collapse and we would be back in the stone age. The moon-shot was a technological achievement; few people have any idea of the enormous amount of basic research which had to be executed to make it possible.

Basic research may seem very expensive. I am a well-paid scientist. My hourly wage is equal to that of a plumber, but sometimes my research remains barren of results for weeks, months or years and my conscience begins to bother me for wasting the taxpayer's money. But in reviewing my life's work, I have to think that the expense was not wasted. Basic research, to which we owe everything, is relatively very cheap when compared to other outlays of a modern society. The other day I made a rough calculation which led me to the conclusion that if one were to add up all the money ever spent by man on basic research, one would find it to be just about equal to the money spent by the Pentagon this past year.

These relationships are still clearer in biomedical research. The cell is a wonderfully complex and precise little mechanism, and a disease is but disorder in this mechanism. The aim of medicine is to prevent or repair such disorders. It is impossible to repair a machine without knowing how it works. The basic research in biology is aimed at an understanding of the cellular machinery. Cancer, too, is but a disturbance in this basic machinery. Everybody has a one-third chance of dying of it. I have lost the two human beings I loved most from cancer. It happens that I am writing these lines one day after it was announced during a news program on my TV screen that the government has cut its support of cancer research by five million dollars, greatly reducing the speed at which this terrible problem can be solved. On the same program the announcer stated that the House had voted 20 billion dollars of additional appropriations to the military, especially for the ABM. I wondered whether the Congress knew that 20 billion is 4 thousand times 5 million? That one person is dying of cancer every two minutes in the United States alone, mostly after suffering through terrific agony? That every lady reading this book has a one-third chance of having a breast removed by an operation which will most likely not save her from agony and death later. Even Congressmen and Senators have a one-third chance of dying of cancer, however powerful they may feel in their present exalted positions. Cancer is a problem which can be solved, and even if we forget all humanistic aspects, this budget cut in cancer research seems to be a poor economy because the treatment of cancer patients costs billions of dollars and the five million dollars cut back could easily have been economized, even by a minor progress in cancer research.

XVII. THE WAY OUT

There is one over-riding question that concerns us all: how can we get out of the fatal groove we are in, the one that is leading towards the brink? There is no way to do so. We must make a new beginning, but it is only the youth who can make it by building a new world. The question, then, is what the principles are on which this new world can be built.

Religion? Unfortunately there is no one religion which embraces all mankind or appeals to all mankind. The Christian religion is too much mixed up with dogmatism and church imperialism. It also has a very poor record. No other religion has caused as much bloodshed and suffering in the world as has Christianity.

It is evident to me that a world created by science can be run safely only by the spirit and methods of the science which created it, and created science itself. It is impossible to create a new world by means of science and then run it on the basis of outdated, sentimental principles like fear, lust for power and domination. Science has two main values to offer which can help build a new world. The one is its spirit and the other its method.

The spirit of science is that of good will, mutual respect and human solidarity. This results from the fact that science was not built by any single nation or race, but is the common property of man, having been created by peoples of the most different backgrounds and descents. Scientists form one single community which knows no borders of space or time. Although I am living in a certain community at a certain time, Newton, Pasteur and Bach

75

are my daily companions. Any scientist is closer to me than my own milkman and, as the Pugwash conference showed, we scientists can discuss our problems peacefully, even if governments would like to see us separated as enemies.

The main value science has to offer for the practical solution of our problems is its method, which created science itself. The essential point about scientific method is that it *meets problems as problems* and searches for the best solution, irrespective of prejudices and chauvinism. We do not ask who is right, but ask what the truth is. Searching for the truth we collect data and analyze them with cool heads, with uncompromising honesty, unbiased by interest or sentiment, fear or hatred. If we have an adversary in our work, we look upon him as an associate with whom, together, we will find the truth and the best solution. We accept no statement without solid evidence, knowing that even governments make false statements. This is the foundation on which science is built and is the only method for building a safe new world, resolving the differences between nations, creating peace without fear, hunger and disease, with undreamt-of wealth, dignity and happiness: a world not based on force but on decency, equity and good will. If this method were applied at the Paris peace talks, the war would soon be over.

Science has no blueprint for a new world; it can offer only its spirit and method for devising such a blueprint.

I am not dreaming of a Utopia, only of a world in which problems are not resolved by force but by intelligence, good will and equity; a world in which killing, no matter the reason, and the destruction of a fellow man's life or home, is a crime; a world in which our youth will not have to spend their best years studying organized manslaughter, in which neither force nor megatons nor poison

76

gases will decide a nation's standing but the sum of its knowledge, its ethics, the gifts it makes to mankind, the happiness it gives to men, the measure in which it lifts human life.

Science even has the kernel of the new religion which the modern mind is longing for. All religions are based on the idea that there are forces greater than our own, that we are not the masters but the fruits of creation. The greatest compliment one can pay to a creative artist is not in words of praise but in the study and appreciation of his creation. If there is a creator, the greatest homage we can pay to him is the study, understanding and appreciation of his creation. This is what science does, and if I am doing my research I have the feeling of performing a divine service and feel as Haydn did about composing: he put on his best clothes whenever he sat down to compose, as though he were in church.

XVIII. ADDITIVITY AND THE BIPARTISAN WORLD

The human mind is complex, ambivalent. Most of us have a side in our mind which likes peace, equity, decency; but we also have a side which prefers simple solutions: terror, violence and brutality.

According to one of the most basic sciences, thermo-dynamics, there are two kinds of qualities: qualities which one can add up are additive, and qualities which one cannot add up are non-additive. You and me, together we can lift twice as big a load as we can one by one. The force of our arms is additive, but our body temperatures are not additive. Your 98°F and my 98°F do not add up to 196°, we remain 98° however close we get together.

Unfortunately, stupidity is additive while intelligence is not. Stupidity likes the crowd, while intelligence seeks its own way. Violence and brutality are not only additive, they tend to unite and have a strong coherence; and by uniting they acquire a momentum which is out of proportion with their real mass. Under government leadership they can become frightening. The only way men of good will can protect themselves is by uniting themselves.

All this has been rendered of vital importance by modern science, which has enabled man to create heaven or hell on earth. We can create a world of happiness, health and dignity, or a world of terror and misery, marching to its doom. The choice is ours.

All of us have a little stupidity in us, say 10 per cent, and this stupidity is additive. This is why governments often appeal to our baser instincts and are sure of the majority of votes when appealing to the lowest common denominator.

What about communism? It is here and here to stay, and democracy is in competition with it, competing for world rule. In the past communism often invoked force and terror as its instrument of establishing itself. This is true, but there can be no doubt that communism has changed a great deal since Stalin passed. It has also matured and gotten over its messianic expansive beginnings. The battle going on now is for the minds of men; the outcome of this battle does not depend on numbers of missiles, but on the question of which system can raise life higher, give more happiness to man and lift the great underdeveloped masses out of their misery.

Western civilization has given much to mankind, but with its selfish colonialism it has also made a poor record. We Americans believe in the bipartisan system. The trouble with the world, until the First World War, was that there was no competition, there was one party only, the party of the Westerner. Now there are two parties: democracy and communism. Why not embark on a noble competition by showing which of the two systems can create a better, freer, happier life.

The Soviet Union has changed a great deal and has made, of late, many overtures to better relations; naturally, we reject these as propaganda. It is impossible to get anywhere by rejecting. The longest journey, one of ten thousand miles, has to begin with the first step. Joining forces with the Soviets, we could make order in this lawless world, instead of furthering lawlessness.

The great hope of mankind is still the United Nations. It has wonderful possibilities but has to be a real "United Nations." That the greatest and oldest of all nations, China, is still not admitted to it is, I think, a sample of our insincerity.

POST SCRIPT

The My Lai massacre came to light as I was about to finish this book. It was very painful to see the good name of my beloved country blackened, the country which for a long time was the flag-bearer of human ideals. We Americans knew or guessed that the sort of thing that happened in My Lai was going on, and we are guilty of even worse mayhem and torture, guilty by association.

What will happen, naturally, is that the main culprits will be court-martialled and receive stiff punishments. The Army will try to polish its tarnished image and will try to show its innocence by shifting the blame to a few individuals, especially to Lieutenant Calley. What frightens me about Calley is not that he allegedly killed, but that, according to witnesses, he is a decent fellow who was a good student and a good soldier—he apparently always did his duty and never revealed any traits of criminality. This frightens me because it shows how terribly brutalizing wars and military life are, how they are capable of turning decent fellows into mass murderers who can shoot women and children down in cold blood. The culprits are those who turned Lt. Calley into a murderer. If I were his judge I would dismiss the case, exonerate Lt. Calley and his fellows, and pass a severe judgment on the society which created the institutions that made murderers out of decent people. After all, the main object of prolonged military training is to teach men to obey orders without questioning. This, it appears, is exactly what Lt. Calley did. He was being a good soldier. There is a Hungarian saying: "The fish always starts to smell at its head."

81

There are other points, too, which disturb me about this case. There were attempts to prove that the massacre could not have taken place because My Lai was wiped out earlier by an air attack. If this could be proven, there would be no problem. What disturbs me is that if this were true, all the women and children would have been killed just the same. Why, then, is the point-blank shooting of civilians on the ground worse than an air attack, which belongs to the routine procedures? Just because the pilot who drops the bombs does not see his victims? Killing the victim by a gun, point-blank, is a more honest and open way of doing it. To see what one is doing and accept the responsibility for it is more honest and manly.

What makes the whole situation so gruesome is that it was we, the United States, who laid down the principle that orders do not excuse misdeeds, and everybody should act according to his own moral standards, being responsible to his own conscience. But if so, where are the limits? My own conscience tells me that it is wrong to go ten thousand miles away from home to kill people in order to keep in power the anti-democratic and corrupt government of a police state. Of course, if I was a young man sent to Vietnam I would kill, but only in self-defense. If I did not have to go there, however, I would not have to defend myself. Thus if I was indeed 20 years old today and was called up, and I followed my conscience, I would certainly tear up my draft card. I would certainly then be sent to jail, maybe for five years of hard labor. The judge who would send me to jail for refusing to kill would be, in principle, the same judge who would condemn Lt. Calley for having obeyed the order to kill.

The whole situation shows what a terrible mess and muddle we are in. The idea of an armed force, or any force, for that matter, as an instrument of policy, is out of date in the 20th century. Entirely new means have to be found for the solution of political problems, which demand good will and intelligence, not force, for their solution.

The number of casualties in the Vietnam war has reached the number of total casualties of World War I and is half of the number of total casualties of World War II (not counting the civilians), and the killing still goes on, tarnishing the good name of my country. Patriotism, as I see it, demands that we all refuse to take part in this process. If there are any who helped to brighten our good name and standing among nations, it is the Moratorium marchers, people who are demonstrating that this war is not the war of the American people but only that of its army and government.

Mankind has reached a crossroads, and is confronted by two road signs pointing in opposite directions. One of these is symbolized by the happenings at My Lai. It points towards a dark world, dominated by military-industrial complexes and conducted by fear, hatred and distrust. Its features are terror and the building of monstrous instruments of murder—atomic bombs and submarines, napalm, fragmentation bombs, defoliants, nerve gases, etc. This road leads to doomsday, to the deserved disappearance of man from his polluted little globe.

The other road sign points in the opposite direction. It would lead man to a sunlit, peaceful and clean world, marked by good will, human solidarity, decency and equity, and free of hunger and disease, with a place for everyone.

Man can waste no more time in making a decision as to which road he is going to take. If he delays much longer, he might as well not have pondered the question in the first place. It seems like such a simple decision. Or is it?

APPENDIX

PSALMUS HUMANUS AND SIX PRAYERS*

Psalmus Humanus

My Lord, Who are You?
Are You my stern Father,
Or are You my loving Mother
In whose womb the Universe was born?
Are You the Universe itself?
Or the Law which rules it?
Have You created life only to wipe it out again?
Are You my maker, or did I shape You,
That I may share my loneliness and shun my responsibility?

God! I don't know who You are
But I am calling to You, for I am in trouble,
Frightened of myself and my fellow men!
You may not understand my words,
But comprehend my wordless sounds.**

---

* I wrote these prayers in the summer of 1964. I subsequently
recorded them in recitative style to musical accompaniment composed
and played by Agi Jambor.
** This refers to the music on the recording.

87

First Prayer: *God*

My Lord!
You are greater than the world You created,
And Your house is the Universe.

I shaped You to my own image
Thinking You vicious, greedy and vain,
Desirous of my praise and sacrifices,
Revengeful of my petty trespasses,
Needful of the houses I build you
While my fellow men I let go without food and shelter.

God! Let me praise You by improving my corner of Your
    Creation
By filling this little world of mine
With light, warmth, good will and happiness.

Second Prayer: The Leaders

My Lord!
We elect leaders to lead us,
And give You servants to serve You.

But the leaders don't lead us to You,
They don't listen to our mute voices of craving for peace,
They are corrupted by power, lead man against man.
And the servants we give You don't serve You,
They serve power and bless our guns,
Torture and kill my fellow men in Your name.

God! Give us leaders who are Your servants,
Who lead us to You, lead us to peace,
Lead man to man.

Third Prayer: The Heart and the Mind

My Lord!
You have given me a heart capable of love and thirsty for
    love,
You have given me a mind capable of clear thought and
    creativeness,

And I have filled my heart with fear and hatred,
And my heart corrupts my mind and makes it build
    monstrous instruments of murder

To destroy Your world, myself and my fellow men,
And damage the sacred stuff life is made of.

God! Clean my heart, lift my mind,
And make me my brother's brother.

Fourth Prayer: Energy and Speed

My Lord!
You have revealed to us the secret energies of matter
To ease our toil and elevate life,
You have taught us to travel faster than the sound we make
That distance should no more separate man from man.

We toil to press these energies into shells
In which to send them to the distant corners of the earth,
To bring misery and destruction to our fellow men,
Leaving the earth scorched and barren of life.

God! Let me not destroy the temple of life,
Let me use my knowledge to my advantage, to elevate life,
Lend dignity to the short span of my existence.

Fifth Prayer: The Earth

My Lord!
You have given us this lovely globe to live on,
Hidden untold treasures in its bowels,
Enabled us to comprehend Your work,
Ease our toil, ban hunger and disease.

We are digging up those treasures to squander them,
To build them into formidable machines of destruction,
With which to destroy what other men have built
Which will turn against me, destroy me and my children.

God! Let us be Your partners in creation
By understanding and improving Your work,
Making this globe of ours a safe home
For wealth, happiness and harmony.

Sixth Prayer: Children

My Lord!
You have separated the sexes that in their mutual search
The deepest chords of our souls may vibrate in the highest
  harmonies.
Out of the search spring our children, lovely children
Who come to us with clean and empty minds.

And I fill these minds with my hatreds, fears and prejudices,
My bomb shelters teach them the darkness of life and the
  futility of endeavor,
And when they grow up, ready for noble deeds,
I make them study organized manslaughter,
Wasting their best years in moral stagnation.

God! Save my children,
Save their minds
That my corruption may not corrupt them,
Save their lives
That the weapons I forge against others may not destroy
  them,
That they may be better than their elders,
That they may build a world of their own,
A world of beauty, decency, harmony, good will and equity,
That peace and love may reign,
For ever.

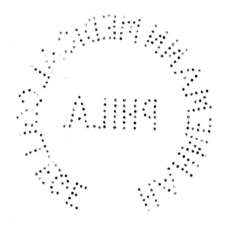